MEDIA GUIDE for St

Chemistry, SIXTH EDITION, WITH VIRTUAL TOOLBOX

Steven S. Zumdahl

Susan A. Zumdahl

Virtual Toolbox
Flexible, integrated learning tools

This handy guide provides information about the media resources available with the Sixth Edition. These tools can help you succeed in your course.

Contents

Printed in the U.S.A.

Media Guide for Students ISBN: 0-618-54406-2

Student Text and Media Guide for Students ISBN: 0-618-54407-0

Visit: **college.hmco.com/PIC/zumdahl6e** For Technical Support: **(800) 732-3223** or **support@hmco.com**

Dear Student,

Welcome to your new adventure in chemistry! You will build on previous experiences with chemistry to develop a broader and deeper understanding of this marvelous and diverse field. Be aware, though, that the important skills you develop will include both the chemistry and the methods you develop to learn it. No one can *teach* you chemistry. You must *learn* the material on your own by using whatever resources are available especially study aids and expert assistance.

Explore the Media Guide carefully to discover which of the many resources best suit your needs and method of study. The **textbook** is your first and most important resource and should always be your starting point to learn course content. The **Student Website** gives you the freedom to access additional study aids such as flashcards, a brief overview of each chapter, and a collection of molecular animations and lab demonstrations. It also offers a molecular library and ACE self-tests to help you prepare for quizzes and exams. In addition, you can use the Keywords list to test your understanding of each word or phrase or to clarify a new definition.

Through **SMARTHINKING**™, you can receive live, online tutoring from qualified instructors during peak study hours. In addition, you can submit questions and get feedback within 24 hours. **Eduspace**® provides another avenue to the resources described above, along with online homework assignments given by your instructor.

C. Weldon Mathews, Ohio State University

SMARTHINKING™ live, online tutoring

Why use SMARTHINKING?

- **Access live help when you are unable to meet with your instructor.**
- **Get personalized tutoring with difficult concepts.**
- **Review previous problem-solving discussions to help prepare for exams.**

This live, online service provides personalized, text-specific tutoring when you need it most. **With *SMARTHINKING* you can**:

- connect immediately to live help during typical study hours: Sunday through Thursday from 2 P.M. to 5 P.M. and 9 P.M. to 1 A.M. EST.
- submit a question to get a response from a qualified e-structor within 24 hours.
- use the whiteboard with full scientific notation and graphics.
- pre-schedule time with an e-structor.
- view past online sessions, questions, or essays in an archive on your personal academic homepage.
- view your tutoring schedule.

E-structors help you with the process of problem-solving rather than supply answers.

▼ Whiteboard

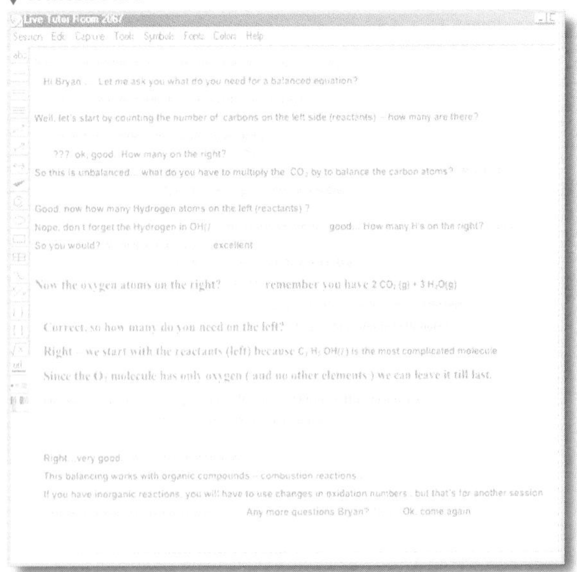

▼ SMARTHINKING™ live, online tutoring homepage

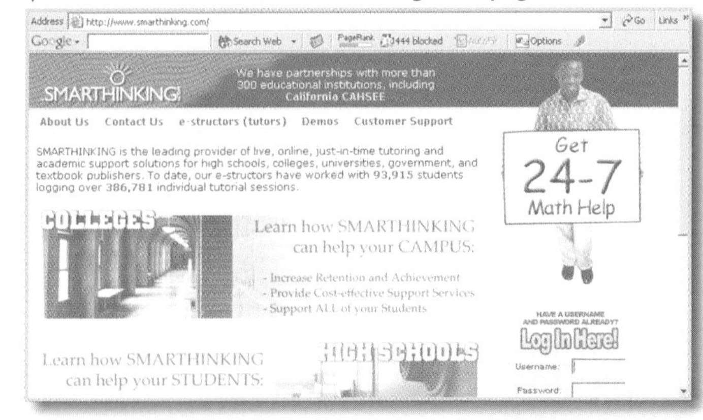

Logging on to SMARTHINKING *live, online tutoring*

1. **Go to smarthinking.com/houghton.html.**
2. **Follow the instructions online to set up your student account using the password located on the inside front cover of this guide.**

For technical support for SMARTHINKING, call (888) 430-7429, extension 1 or email info@smarthinking.com.

If your instructor asked you to register for *Eduspace,* you can also access *SMARTHINKING* directly from within *Eduspace* once you have logged in.

Student Website *Chemistry,* Sixth Edition

Why use the Student Website?

- **Get help understanding core concepts at any time.**
- **Visualize molecular-level interactions.**
- **Practice problem solving with ACE practice tests to prepare for quizzes and exams.**

Organized by chapter, the *Student Website* supports the goals of the Sixth Edition with the following resources:

Visualization, Practice, and Tutorial Tools:

- Understanding Concepts and Visualizations tutorials that use images, animations, and videos to help reinforce core chemical concepts
- Quiz questions that test your knowledge of the Visualizations, animations, and videos
- Flashcards of key terms and concepts
- Houghton Mifflin's ACE practice tests
- Molecule library

Additional Resources and Study Aids:

- An interactive periodic table
- Glossary from the text
- Link to SMARTHINKING online tutoring
- Information on careers in chemistry

Interactive flashcards

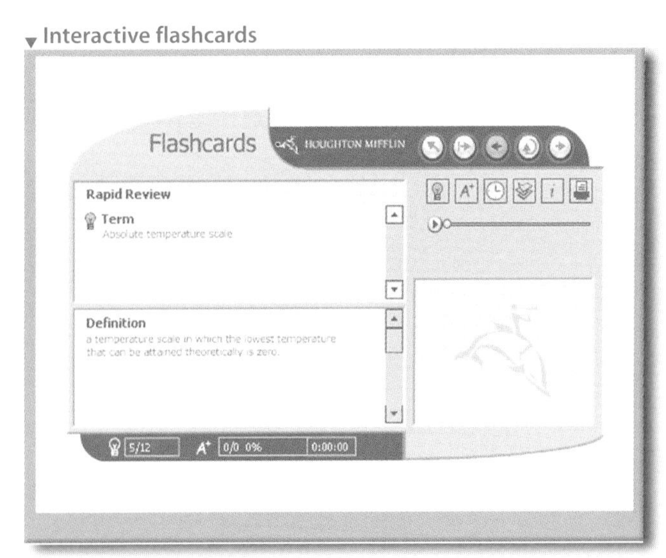

ACE self-quizzes

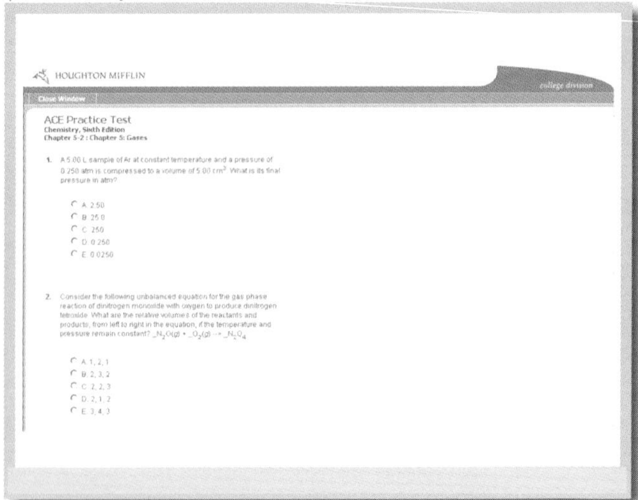

Student Website *Chemistry,* Sixth Edition

Logging on to the Student Website

1. **Go to college.hmco.com/pic/zumdahl6e.**
2. **Select "Students."**
3. **Enter the user name and password located on the inside front cover of this guide.**

If your instructor asked you to register for *Eduspace*, you can also access the *Student Website* directly from within *Eduspace* once you have logged in.

Eduspace® (powered by Blackboard™)

Why use Eduspace?

- Complete assignments specifically chosen by your instructor.
- Practice problem solving with online homework to improve your scores on quizzes and exams.
- Access live, online tutoring through SMARTHINKING anytime you need extra help.
- Visualize chemical concepts at the molecular level.
- Extend your understanding of chapter concepts.

Online Homework

Accessible anytime you need it, *Eduspace* includes an automatically graded online homework system featuring *ChemWork™*—unique assignments that help you learn the process of thinking like a chemist. As you work through text-based exercises, a system of interactive hints is available to help you think through each problem. In addition, you can work on algorithmic end-of-chapter problems which include helpful links to examples, art, and tables in the text itself to reinforce your understanding of key concepts. Using *Eduspace,* you can practice problem solving and check your progress in the gradebook anytime.

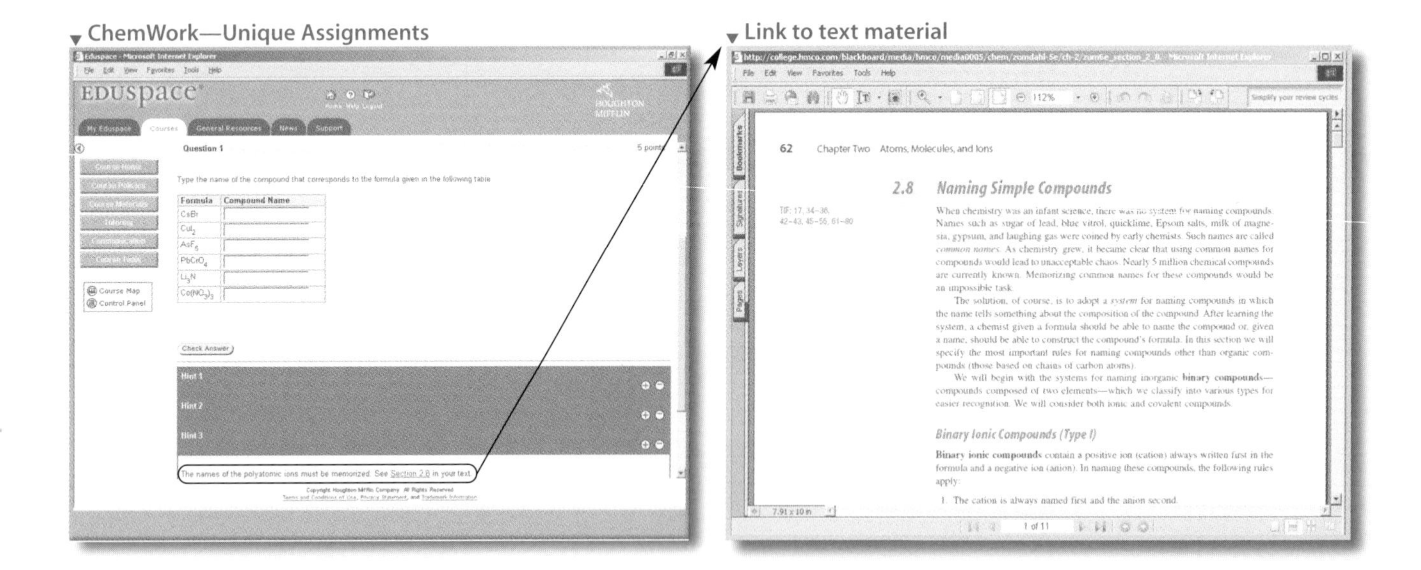

Eduspace® (powered by Blackboard™)

Logging on to Eduspace (powered by Blackboard)

1. **Go to www.eduspace.com.**
2. **Register for Eduspace.**

If your instructor is using the *Eduspace* program, refer to your *Eduspace* Getting Started Guide for Students for information and access.

For your reference

Periodic Table of the Elements

1 1A	2 2A	3	4	5	6	7	8	9	10	11	12	13 3A	14 4A	15 5A	16 6A	17 7A	18 8A
1 **H** 1.008																	2 **He** 4.003
3 **Li** 6.941	4 **Be** 9.012											5 **B** 10.81	6 **C** 12.01	7 **N** 14.01	8 **O** 16.00	9 **F** 19.00	10 **Ne** 20.18
11 **Na** 22.99	12 **Mg** 24.31											13 **Al** 26.98	14 **Si** 28.09	15 **P** 30.97	16 **S** 32.07	17 **Cl** 35.45	18 **Ar** 39.95
19 **K** 39.10	20 **Ca** 40.08	21 **Sc** 44.96	22 **Ti** 47.88	23 **V** 50.94	24 **Cr** 52.00	25 **Mn** 54.94	26 **Fe** 55.85	27 **Co** 58.93	28 **Ni** 58.69	29 **Cu** 63.55	30 **Zn** 65.38	31 **Ga** 69.72	32 **Ge** 72.59	33 **As** 74.92	34 **Se** 78.96	35 **Br** 79.90	36 **Kr** 83.80
37 **Rb** 85.47	38 **Sr** 87.62	39 **Y** 88.91	40 **Zr** 91.22	41 **Nb** 92.91	42 **Mo** 95.94	43 **Tc** (98)	44 **Ru** 101.1	45 **Rh** 102.9	46 **Pd** 106.4	47 **Ag** 107.9	48 **Cd** 112.4	49 **In** 114.8	50 **Sn** 118.7	51 **Sb** 121.8	52 **Te** 127.6	53 **I** 126.9	54 **Xe** 131.3
55 **Cs** 132.9	56 **Ba** 137.3	57 **La*** 138.9	72 **Hf** 178.5	73 **Ta** 180.9	74 **W** 183.9	75 **Re** 186.2	76 **Os** 190.2	77 **Ir** 192.2	78 **Pt** 195.1	79 **Au** 197.0	80 **Hg** 200.6	81 **Tl** 204.4	82 **Pb** 207.2	83 **Bi** 209.0	84 **Po** (209)	85 **At** (210)	86 **Rn** (222)
87 **Fr** (223)	88 **Ra** 226	89 **Ac**† (227)	104 **Rf** (261)	105 **Db** (262)	106 **Sg** (263)	107 **Bh** (264)	108 **Hs** (265)	109 **Mt** (268)	110 **Ds** (281)	111 **Uuu**	112 **Uub**		114 **Uuq**				

Alkali metals: group 1 · Alkaline earth metals: group 2 · Transition metals: groups 3–12 · Halogens: group 17 · Noble gases: group 18 · metals ← → nonmetals

*Lanthanides	58 **Ce** 140.1	59 **Pr** 140.9	60 **Nd** 144.2	61 **Pm** (145)	62 **Sm** 150.4	63 **Eu** 152.0	64 **Gd** 157.3	65 **Tb** 158.9	66 **Dy** 162.5	67 **Ho** 164.9	68 **Er** 167.3	69 **Tm** 168.9	70 **Yb** 173.0	71 **Lu** 175.0
†Actinides	90 **Th** 232.0	91 **Pa** (231)	92 **U** 238.0	93 **Np** (237)	94 **Pu** (244)	95 **Am** (243)	96 **Cm** (247)	97 **Bk** (247)	98 **Cf** (251)	99 **Es** (252)	100 **Fm** (257)	101 **Md** (258)	102 **No** (259)	103 **Lr** (260)

Group numbers 1–18 represent the system recommended by the International Union of Pure and Applied Chemistry.

Table of Atomic Masses

For your reference

Table of Atomic Masses*

Element	Symbol	Atomic Number	Atomic Mass	Element	Symbol	Atomic Number	Atomic Mass	Element	Symbol	Atomic Number	Atomic Mass
Actinium	Ac	89	(227)†	Gold	Au	79	197.0	Praseodymium	Pr	59	140.9
Aluminum	Al	13	26.98	Hafnium	Hf	72	178.5	Promethium	Pm	61	(145)
Americium	Am	95	(243)	Hassium	Hs	108	(265)	Protactinium	Pa	91	(231)
Antimony	Sb	51	121.8	Helium	He	2	4.003	Radium	Ra	88	226
Argon	Ar	18	39.95	Holmium	Ho	67	164.9	Radon	Rn	86	(222)
Arsenic	As	33	74.92	Hydrogen	H	1	1.008	Rhenium	Re	75	186.2
Astatine	At	85	(210)	Indium	In	49	114.8	Rhodium	Rh	45	102.9
Barium	Ba	56	137.3	Iodine	I	53	126.9	Rubidium	Rb	37	85.47
Berkelium	Bk	97	(247)	Iridium	Ir	77	192.2	Ruthenium	Ru	44	101.1
Beryllium	Be	4	9.012	Iron	Fe	26	55.85	Rutherfordium	Rf	104	(261)
Bismuth	Bi	83	209.0	Krypton	Kr	36	83.80	Samarium	Sm	62	150.4
Bohrium	Bh	107	(264)	Lanthanum	La	57	138.9	Scandium	Sc	21	44.96
Boron	B	5	10.81	Lawrencium	Lr	103	(260)	Seaborgium	Sg	106	(263)
Bromine	Br	35	79.90	Lead	Pb	82	207.2	Selenium	Se	34	78.96
Cadmium	Cd	48	112.4	Lithium	Li	3	6.941	Silicon	Si	14	28.09
Calcium	Ca	20	40.08	Lutetium	Lu	71	175.0	Silver	Ag	47	107.9
Californium	Cf	98	(251)	Magnesium	Mg	12	24.31	Sodium	Na	11	22.99
Carbon	C	6	12.01	Manganese	Mn	25	54.94	Strontium	Sr	38	87.62
Cerium	Ce	58	140.1	Meitnerium	Mt	109	(268)	Sulfur	S	16	32.07
Cesium	Cs	55	132.9	Mendelevium	Md	101	(258)	Tantalum	Ta	73	180.9
Chlorine	Cl	17	35.45	Mercury	Hg	80	200.6	Technetium	Tc	43	(98)
Chromium	Cr	24	52.00	Molybdenum	Mo	42	95.94	Tellurium	Te	52	127.6
Cobalt	Co	27	58.93	Neodymium	Nd	60	144.2	Terbium	Tb	65	158.9
Copper	Cu	29	63.55	Neon	Ne	10	20.18	Thallium	Tl	81	204.4
Curium	Cm	96	(247)	Neptunium	Np	93	(237)	Thorium	Th	90	232.0
Darmstadtium	Ds	110	(281)	Nickel	Ni	28	58.69	Thulium	Tm	69	168.9
Dubnium	Db	105	(262)	Niobium	Nb	41	92.91	Tin	Sn	50	118.7
Dysprosium	Dy	66	162.5	Nitrogen	N	7	14.01	Titanium	Ti	22	47.88
Einsteinium	Es	99	(252)	Nobelium	No	102	(259)	Tungsten	W	74	183.9
Erbium	Er	68	167.3	Osmium	Os	76	190.2	Uranium	U	92	238.0
Europium	Eu	63	152.0	Oxygen	O	8	16.00	Vanadium	V	23	50.94
Fermium	Fm	100	(257)	Palladium	Pd	46	106.4	Xenon	Xe	54	131.3
Fluorine	F	9	19.00	Phosphorus	P	15	30.97	Ytterbium	Yb	70	173.0
Francium	Fr	87	(223)	Platinum	Pt	78	195.1	Yttrium	Y	39	88.91
Gadolinium	Gd	64	157.3	Plutonium	Pu	94	(244)	Zinc	Zn	30	65.38
Gallium	Ga	31	69.72	Polonium	Po	84	(209)	Zirconium	Zr	40	91.22
Germanium	Ge	32	72.59	Potassium	K	19	(39.10				

*The values given here are to four significant figures where possible.

†A value given in parentheses denotes the mass of the longest-lived isotope.

Print Supplements Available Separately

Partial Solutions Guide (ISBN 0-618-22163-8)

Thomas J. Hummel, *University of Illinois, Urbana*

While the text provides alternate strategies for students to tackle difficult chemistry equations, the Solutions Guide offers in-depth solutions to the text problems. This Partial Solutions Guide contains solutions to all of the odd-numbered end-of-chapter problems.

Study Guide (ISBN 0-618-22162-X)

Paul B. Kelter, *University of Illinois*

The Study Guide reflects the unique problem-solving approach of the book. Minor revisions to the Sixth Edition Study Guide include new, worked-out examples and material on the new Spectroscopy section in the text. Furthermore, the Study Guide contains summaries of chapter sections, exercises, self-tests, and answers to the exercises and self-tests—all designed to help students achieve mastery of the material in each chapter.

To purchase these supplements, ask at your bookstore, visit our website at college.hmco.com, *or call Houghton Mifflin Customer Service at (800) 225-1464.*

System Requirements

Eduspace® System Requirements

Minimum Requirements

- Microsoft Windows 2000 or Windows XP
  or
  Macintosh OS* 10.1, 10.2, or 10.3
- Microsoft Internet Explorer 5.5 (or above)
  or
  Netscape Navigator 6.2 (or higher, 7.1 or higher for Windows XP or Macintosh OS 10.2)
- Sun Java Run-time Environment (JRE) 1.4.x in your web browser (Note: Use of Microsoft JVM is **not supported.**)
- Acrobat Reader 4.0
- Display with 256 colors and 800 x 600 resolution or higher
- 56 K modem connection

Strongly Recommended Hardware/Software

- Microsoft 2000 with Microsoft Internet Explorer 6.0 or Netscape Navigator 7.1**
  or
  Microsoft Windows XP with Internet Explorer 6.0 or Netscape Navigator 7.1**
  or
  Macintosh OS 10.3 with Safari 1.2 or Netscape Navigator 7.0 (or higher)**
- Adobe Acrobat 6.0
- Display with thousands of colors and 1024 x 768 resolution or higher
- T1 line, cable modem, DSL, or company Internet LAN

*Use of Macintosh OS 10.1 or higher is required for users of Math and Chemistry courses. Users of other courses may be able to successfully use Macintosh OS 9.2 with their Eduspace courses.

**Users of Math and Chemistry courses are strongly advised to use Microsoft Internet Explorer or Safari to ensure that symbols in exercises and test questions appear properly.

Depending on your course content, you may need one or more of the following plug-ins: Flash Player 7 or higher, QuickTime 5.0 (6 recommended) RealPlayer 9.0 (10.0 recommended) Shockwave 8.5.1.

What if I have Windows 98, Windows NT or AOL?
Although Houghton Mifflin has not performed certification of Eduspace on Windows 98,Windows NT, or AOL and therefore cannot say that we officially support those platforms, Houghton Mifflin is aware of users who are successfully using Eduspace in those environments. If you encounter problems, once you are online with AOL, start an additional browser such as Microsoft Internet Explorer and access Eduspace using that browser. Note that when using any browser with Eduspace, cookies must be enabled.